Endorsements for "Darwin's Replacement" (First Edition)

"The need for a superintelligent force to create and sustain living things is well set out and without question. I also appreciated all the research that was done to demonstrate the historical importance and recognition of God in the nation lives of the four nations you selected." -- George Matzko, PhD

"*Darwin's Replacement* is a well-written review of the enormous complexity of all life from the perspective of the molecular foundation. The complexity is more than amazing. Most people enjoy learning about amazing feats, and thus the popularity of 'Ripley's Believe It or Not' and similar books. Our body and its complexity is familiar to all and works so well for most of us that we often take it for granted. Mr. Rogers' book helps us to realize that we are all walking miracles and understanding how it functions is both awe-inspiring and helps us to appreciate the body we all live in while on this earth." -- Jerry Bergman, PhD

"*Darwin's Replacement* provides a credible assessment of the weaknesses within mainstream evolution theory and proposes a reasonable, evidence-based alternative for the origin and development of life." -- Nicholas Comninellis, MD MPH

"ATOMIC BIOLOGY promises to restore the true foundations of science back to the realm of observation. Current forays into metaphysical speculation and presupposition by 'experts' seem to have caused division, confusion, and misunderstanding in the scientific conversation." -- Jack Taylor, PhD

"My recommendation would be to rework the book as a series of 1-2 page studies for adult Sunday school classes and/or Christian High School classes. It could be useful as an introduction to the topics, in that format." -- David Snoke, PhD

"Your book demonstrates the amazing complexity of life, starting with even the simplest cell, and the numerous conditions needed to sustain life. That all this could be the result of blind, random evolution is highly implausible, and statistically virtually impossible. Hence, as your book concludes, this points to a superintelligent Creator. Your book also notes that the USA, the UK, Canada, and Australia were all founded on submission to the Christian God, and urges those countries return to acknowledging God, also in science classrooms.

I heartily agree with all this." -- John Byl, PhD

"There are only two possibilities for the existence of life: accidental or purposeful. Using science and mathematics, Atomic Biology proves beyond a shadow of doubt that life cannot be accidental. Then the book shows that the only being capable of the creation of life and its orchestrated maintenance is the historical Omniscient, Omnipotent, and Omnipresent Triune God of the Bible and our Nation." -- Sharon E. Cargo, DVM

"Hi, (Reality R&D). I recently bought the book and am reading it slowly and out loud to myself so that it sticks, but can I say that when I heard Mr. Rogers on Vision I knew that this was a book I'd been waiting for a very long time. I honestly can't put into words how exciting this is for me to finally have something I can refer to when discussing creation and not sound like a loony."
-- Linda Houston, Australia

DARWIN'S REPLACEMENT SERIES - FOR EVERYONE

GOD IS IN THE GROCERY STORE

"THE TRUTH FOR LIFE EDUCATION PROJECT"

LRG

LIFETIME REFERENCE GUIDES INC.
P.O.Box 51613 RPO Park Royal
West Vancouver, BC, Canada V7T 2X9
www.lifetimereferenceguides.com

Copyright © 2024 by Lifetime Reference Guides, Inc.
Cover design by Arneeon.com with Shutterstock image by Olga Lyubkina.

All rights reserved.

No part of this publication may be reproduced in any form by any means, electronic or mechanical, including photocopying, recording, information browsing, storage, or any retrieval system, without specific written permission from the publisher.

ISBNs
978-1-7383082-4-8 (eBook)
978-1-7383082-3-1 (Paperback)
978-1-7383082-2-4 (Hardcover)

Printed in Canada, USA, UK, and Australia

Creator
Biology
Education
Government

Dedication

To Our Superintelligent Creator,
The God of the Governments in Our Focus Nations,
We Bring Praise and Glory and 100% of the Credit
For Creating Our Grown Foods and Us.

Acknowledgements

Our respect and gratitude go out to the forty-five scholars whose input over three decades has helped in the discovery and development of the God-based life-science of "Atomic Biology," nicknamed "God's Biology." Our Creator has probably been using this science (and all His other sciences) since the beginning.

We are also grateful for our editor, David V. Bassett, MSc, and for the encouragers, endorsers, and supporters who help us to keep revealing God's enormous caring works for every person every second of every day.

A Gift For Others

"Always be prepared to give an answer to everyone who asks you to give the reason for the hope (belief) that you have. But do this with gentleness and respect ..." 1 Peter 3:15

This booklet can help you to clearly enhance, protect, and share your belief.

May God help you count all your blessings and share them in a fruitful way.

Contents:

Dedication 4
Acknowledgements 4
Introduction 5
Chapter 1: God Is In The Grocery Store? 7
Chapter 2: Why Does Life Take "Superintelligence"? 17
Chapter 3: Why Not "Evolution"? 21
Chapter 4: God In Our Governments 27
Chapter 5: Why Is Evolution Still Being Taught? 31
Chapter 6: Why Does Our Creator Care For Us? 35
Chapter 7: How Can We Individuals Use These
              New Facts Of Life? 39
Chapter 8: Choices and Consequences 41
About The Authors: 43
Index: 47
Contact Us: 49
Poem 51

Introduction

This book is written to provide solid, verifiable evidence that WITHOUT GOD, THERE WOULD BE NO FOOD AND NO LIFE. This Truth For Life has been purposefully missing from our public education system for about six decades. Teachers and professors have been forced to teach information about Life that many disagree with, but currently cannot change. The teaching that is enforced at this time is that all living things descended from an (unsubstantiated) "common ancestor" that gradually "evolved" into all other entities by "natural selection." There was no intelligence involved and God is purposefully excluded.

This teaching is heavily enforced, to the point that teachers and professors have been fired for daring to mention that there may be some intelligence involved in the construction of living cells and entities. Even courts are used in enforcement suits.

This booklet is written to provide everyone with solid, undeniable reasons for their belief in our Creator, God. It provides clear, logical, obvious, and verifiable evidence showing that enormous, superintelligent physical works and care are absolutely essential for the construction of all living cells and entities.

Darwin's theory is now *factually falsified* and this taught falsehood over the last six decades seems to be a Satan-based deception used to separate our students and society from their Creator.

This booklet is written to bring verifiable "Truth For Life Education" to everyone, including teachers and students, so they can understand how immensely we are all cared for and can feel gratitude for the One who reliably creates our foods from dust, makes us from our foods, and gives us Life.

Does it seem logical that the greatest Wisdom for us will come from the greatest Intelligence known to mankind?

We can observe His awesome works in progress just by planting vegetable seeds in soil and adding His water and sunshine.

The green shoots and whole vegetables that appear, don't "just happen." Every complex cell part is carefully constructed using the correct numbers of the correct atoms from the soil and water.

Three decades of research with input from forty-five scholars, helped this summary booklet to show:

1. The absolute essentiality for Superintelligence and constant care, far above the human level, for building and sustaining all living cell parts, cells, and entities, including us;
2. It will show the amazing speed of the superintelligent works performed for us in our bodies every second of every day;
3. It will also show the phenomenally careful precision of the brilliant construction works essentially used in building each of our cell-parts, cells, and the total "us."

These are all superintelligent works that "evolution" is incapable of performing because it has no intelligence to use.

Our Creator shows His immense love and care for each one of us every second of every day as you will see in this booklet.

Because of His high recognition by our governments, **our students have the unalienable right to be taught *"Why God Is So Highly Recognized By Their Governments."*** See Chapter 4.

Chapter 1: God Is In The Grocery Store

What? Where is He?

He is actually everywhere in every atom because He has to provide a constant supply of energy to keep the electrons moving forever, BUT, when we go into a grocery store, what do we see? If it is a big store, we see display after display of delicious fruits, vegetables, meats, fish, cereals, milk, cheese, sugars, and more. Those great foods are all constantly and reliably produced for our Life. This takes superintelligence, enormous work, and care.

Are they important for our Lives? You know they are. Where would we be without food?

Olga Lyubkina/Shutterstock

But who did all the amazing work to design them and reliably create them out of dust for our life?

DON'T BE SURPRISED if you begin to hear words like,

"Thank you, Lord!" and "Praise the Lord!" as you shop around your grocery store. People are beginning to understand and appreciate that it is only our Creator who constantly and reliably creates all those delicious and nutritious grown foods for our Life. And all the superintelligent works to make all our grown foods out of dust, He provides to us at no charge for His work. Then He makes every one of our trillions of living cells out of our foods. All these works for us are **gifts** just because He loves each one of us and cares for us that immensely, regardless of our race, color, creed, or belief.

He gives us a whole lifetime to learn to appreciate Him and His caring works for us, but He is understandably disappointed if we show no gratitude.

Of course, the farmers, the truckers, and the grocery store staff have to be paid for their work, but God makes no charge.

How do we know that it is God who does all this work for us and that it is not done by "evolution"? Here is how: Over the last seventy-plus years, scientists have been trying to make a living cell out of elements. However, even with our vast accumulation of scientific knowledge and highly sophisticated equipment, we cannot come anywhere close. We simply do not have enough intelligence. Therefore, greater intelligence, far above mankind's level, *is essential* for building living cells and entities out of elements. We call it "Superintelligence."

There is only one such brilliant and caring entity known to mankind. Our governments and the majority of our citizens call this entity "God."

"Darwinian evolution," as taught in our public schools, claims there is no god or intelligence involved in building living cells. That theory is now *factually falsified* as the explanation for both the origin and the cause of Life. There is no "common ancestor" and there are no living cells produced by "natural selection."

Fortunately, even Richard Dawkins, the eloquent spokesman for many evolutionists, now considers himself a "cultural Christian."[1] Hopefully, this feeling will spread to his followers as the Truth of Life sinks into their honest understanding.

Our national founders had it right when they honored the life works of our Creator highly and formally with our national holidays of Christmas, Easter, and Thanksgiving Day, in our Declarations, Pledges of Allegiance, National Mottos, Justice Systems, Constitutions, National Anthems, on War Memorials, Public Buildings, Currencies, and more with slight variations, nation to nation. Our focus nations are USA, UK, Canada, and Australia.

Our students have the unalienable right to be taiught "Why God Is So Highly Recognized By Their Governments."

Let's clarify two common stumbling blocks in education:

(1) For those who think we cannot teach publicly about the God of our nation because of a (mis)conception regarding "separation of church and state," please understand why "God" is NOT "The Church." "Churches" are buildings or groups of people who may be Biblical Christians, like Catholics and Protestants, or modified Christians like Mormons, Jehovah's Witnesses, and New Agers, or they may be atheists like Satanists and hundreds of cults that are "churches." But "God" is definitely NOT "The Church." *No "church" can create any living thing.*

God is highly respected for great reasons by many churches, just as He is highly respected by our national governments, as outlined above. See Chapter 4.

The point is, regarding the "separation of church and state," that is a good idea as originally intended, i.e. The church must not dictate what the government is to do and the government must not dictate what the church is to do. But please do not confuse "God" with "the church." They are not the same. As mentioned, the church cannot create any living thing.

(2) Let's also clarify our understanding of the most influential troublemaker on Planet Earth, Public Enemy #1, Satan himself. He has deceived so many undiscerning individuals and persuaded them to try to remove God from our society. They do not realize that God is their Creator, Sustainer, and Maintainer of their Life and without Him, they would not have any life at all. He is the only one capable of making our grown foods out of dust and water, plus making our cells out of our foods. He deserves great praise and thanksgiving, not scorn or ignorance for His constant, immense, caring works for everyone every second of every day.

When God said in His Word, the Holy Bible, "For dust you are and to dust you will return," He was not 'joking.'

The basics are simple enough for a fifth-grader, like this:
1. As many of us learned in grade five, "Atoms are the building blocks of universe matter," and that includes us.
2. The atoms for us have to come mainly from our foods.
3. The atoms for our foods have to come mainly from the dust (soil elements) of gardens, fields, and orchards.
4. So, most of the atoms for our foods and us come from the dust. Isn't that logical and easily clear?
5. But how do we know that God does the assembly work? This has been debated for over 160 years!!!
6. For those who want the Truth, it turns out that the answer is now very clear, simple, obvious, and verified, like this: Scientists have been trying to build a living cell from elements for over 70 years. We cannot come anywhere close because our living cells are so phenomenally complex. Even evolutionists are moving away from Darwin's theory because of this complexity.
7. The problem is that we scientists do not have anywhere near enough intelligence to build a single living cell from elements, even in highly sophisticated labs with the best equipment and our vast accumulation of knowledge.
8. Since mankind does not have enough intelligence to build even the simplest living cell out of elements, therefore, obviously, far greater intelligence is essential; we call it "Superintelligence." As mentioned before, there is only one entity known to mankind to have such Superintelligence and that is our Creator, the God of our nations, in whom we can trust.

We now call the Theory of Evolution "a factually falsified theory for both the origin and the cause of life."

Before waking up to this basic fact of Life that mankind does not have enough intelligence to build even one living cell, we noted there was a related breakthrough in 2016. The three Nobel Prize winners in Chemistry, a Frenchman, Jean-Pierre Sauvage, an Englishman, Sir John Fraser Stoddart, and a Dutchman, Bernard L. Feringa, won this prestigious prize as a group after 33 years of research and development to build a few simplistic molecular machines. These are almost infinitely more simplistic than the simplest of the 40+ various molecular machines built for our cells every day of the week. And their molecular machines have NO Life. They have to be stimulated to move with ultra-violet light or something similar. This also proved that mankind does not have enough intelligence to build living cells.

Isn't it interesting that what our forefathers knew by intuition, observation, common sense, logic, and believing God's Word regarding the origin and cause of life, is now proven true scientifically. When they established our nations (USA, Canada, and Australia), including Thanksgiving Days plus Christmas and Easter national holidays, this was part of their high praise and recognition of our Creator, God.

Since intelligent scientists have tried for over 70 years to build a living cell out of elements, what have they accomplished?

The best-known group working on this challenge today is the Craig Venter Group and the best they can do is to copy a small piece of an existing cell and transplant it into another existing (God-built) living cell. They do honestly admit that they can NOT make a living cell from elements.

If any lab group ever could build a cell with all its molecular machinery, they would then have the major challenges of (1) doing this brilliant work within the confines of a microscopic cell membrane that they had built first, then (2) adding that divine breath of life to the non-living atoms, plus (3) doing all this very quickly.

Charles Darwin theorized in his 1859 book, *On the Origin of Species,* that a common ancestor existed for all living things. However, to this day there has been no plausible explanation for how that necessarily complex original ancestor came into existence without intelligent works with atoms. The construction of even the simplest cell is now known to be so complex that a growing number of evolutionists disagree with Darwin's theory.

The following is a simplified image of a cell just to give a bit of an idea of some of the complex parts each of our cells needs to have made for it out of atoms in order for it to function. Atoms do not have Life, so the divine "breath of life" has to be added to each cell part.

ANATOMY OF A CELL

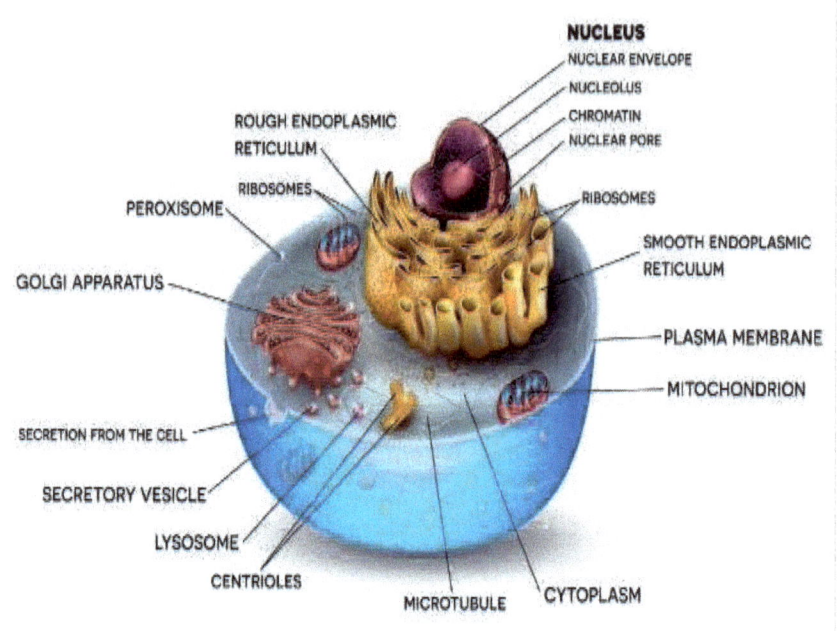

Tefi/Shutterstock

This may look complex, and it is. That's why it takes our superintelligent Creator to design and build all our food cells and our cells which have many similarities.

The brilliant design and construction of cell parts, cells, and living entities like foods and humans have compelled scientists to abandon their former belief that "evolution" created all life.

"Evolution" is a theoretical process that has no intelligence.

Here are some of the many amazing factors that illustrate the brilliant engineering, design, and construction works for life by our superintelligent and caring Creator:

* A 150-pound male is built with about 100 trillion constructed cells including over 200 cell types, and most of these have up to 40

complex molecular machines built within them;
* Each one of our approximately 20 trillion red blood cells (RBCs) is constructed with about 280 million hemoglobin molecules,[2] and each of those molecules is built with about 10,000 correct atoms.[3] So, each one of our red blood cells consists of about 2,800 billion (or 2.8 trillion) correctly selected, counted, and assembled atoms;
* About 2.3 million new red blood cells are produced every second,[4] 24/7, to replace those that are worn-out in this average 150-pound male (i.e. about 6400 quadrillion correct atoms per second). You can estimate the number of atoms **per second** that are being selected and assembled to make **your** red blood cells, using your relative weight divided by 150 times 6,400,000,000,000,000. The constant replacement of our blood cells and almost every other cell type in our body is an enormous and complex task performed for every human every second of every day by our loving, caring Creator.
* Our many cell parts must be carefully constructed to exact specifications. For example, the four DNA bases made for all life forms are constructed precisely to the following formulae:

Adenine (A) - chemical formula: $C_5H_5N_5$
Guanine (G) -   "   " $C_5H_5N_5O_1$
Cytosine (C) -   "   " $C_4H_5N_3O_1$
Thymine (T) -   "   " $C_5H_6N_2O_2$

Notice how carefully the atoms must be selected, counted, and assembled. One atom off is likely lethal. For adenine, $C_5H_5N_5$ is required. $C_5H_5N_6$ will not work, nor will $F_5H_5N_6$ or any other mix.

The required atoms are picked up from foods in our digestive system and carried in our bloodstream to where they are needed for constructing each of our new cell parts. For our DNA, over three billion pairs of these bases must be precisely assembled and arranged in various special sequences (codes) to construct the hardware and assemble the functioning DNA and RNA software instructions required to help operate up to 40 different molecular machines in each one of most of our 200+ different cell types.

How about our amazing eyesight? Even Mr. Darwin said, *"To suppose that the eye with all its inimitable contrivances for adjusting the focus to different distances, for admitting different amounts of light, and for the correction of spherical and chromatic aberration, could have been formed by natural selection, seems, I freely confess, absurd in the highest degree."* [5]

Then he wrongly speculated on how it might just happen without intelligent guidance.

Human Eye Anatomy

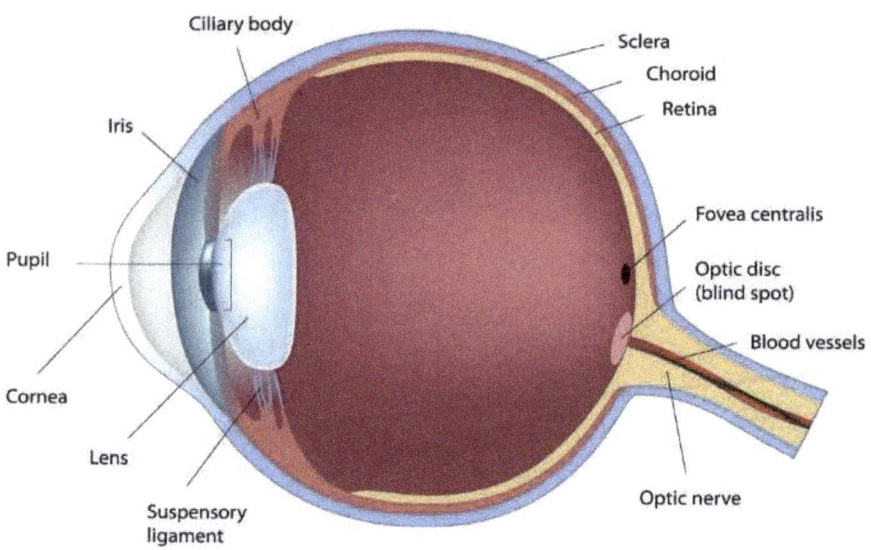

Alila Medical Media/Shutterstock

Then there are all our other amazing senses including hearing, tasting, smelling, and touching (awesome and reliable but so easy to take for granted).

How about our incredible brains? Professor Paul Reber estimates our brain storage capacity is equivalent to about *three million hours of recorded TV shows.*[6] Yet evolutionists claim that our brain is constructed without any intelligent work or care. Seriously?

Questions:

1. What are two of the basic facts that have compelled scientists to move away from belief that "evolution" is the cause of life?
   _____
   _____

2. How many different types of cells have to be made in creating us? _____

3. About how many correctly selected, counted, and assembled atoms does it take to build each one of our red blood cells?
   _____

4. Because it takes about 6400 quadrillion correct atoms every second to make new red blood cells for a 150-pound person, about how many correct atoms are needed to make **your** new red blood cells? (xxx/150 lbs. x 6400 quadrillion) _____ correct atoms every second, 24/7. Thanks to God? _____

5. To make the four bases for our DNA, what elemental atoms are needed? _____
   _____

6. Do these atoms need to be carefully selected, counted, and assembled? _____

7. Where do these atoms come from? _____
   _____

8. In the diagram of a human eye, how many different eye parts are shown? _____

9. Besides seeing, what are the other four senses most people are blessed with? _____
   _____

10. About how many hours of recorded TV shows is our brain capable of storing? _____

11. Which explanation for the cause of life do you believe in ("superintelligent Creator" or "evolution") and why?

_____
_____
_____

12. Do you think it is important to be taught what is true and accurate or does this not matter to you, and why?

_____
_____

References:

[1] Hobson, T., *Is Richard Dawkins A Christian?* spectator.co.uk, 2 April, 2024.

[2] Tortora, G. J., *Principles of Anatomy and Physiology,* John Wiley & Sons, New York, NY, 2008.

[3] Perutz, Max, *Science Is Not a Quiet Life: Unraveling the Atomic Mechanism of Hemoglobin,* World Scientific, Hackensack, NJ, 1997.

[4] Pallister, C. J., *Haematology: Biomedical Science Explained,* Butterworth-Heinemann, Burlington, MA, 1999.

[5] Darwin, Charles, *On the Origin of Species by Means of Natural Selection,* 1st edition, 1859, p. 186, John Murray, London, UK, available online from Darwin-online.org.uk.

[6] Reber, Paul, "What Is the Memory Capacity of the Human Brain," *Scientific American,* May/June 2010, www.scientificamerican.com/article/what-is-the-memory-capacity/.

Chapter 2: Why Does It Take "Superintelligence"?

As mentioned on page 10, in 2016 the three Nobel Prize winners in Chemistry, J.P.Sauvage, Sir J.F.Stoddart, and B.L.Feringa, showed, unintentionally, that mankind has nowhere near enough intelligence and skills to build even the simplest living molecular machine made in our cells. They received this prestigious prize after 33 years of working to make a few simple, nonliving molecular machines. The best they could make were, relatively speaking, almost infinitely more simplistic than the simplest machines built for our new cells every day.

These are only three of the many bright scientists who have tried unsuccessfully over the last seventy years, to make living cells out of elements. This verifies that "Superintelligence," far greater than mankind's level, is essential for building the molecular machines for our cell parts, cells, and ultimately the total *us*.

The evolutionists who say that "natural mechanisms" do this work, invite a much larger problem for their theory: these "natural mechanisms" would first have to be constructed with far more intelligence and better equipment than the Nobel Prize winners had and work them within the microscopic confines of a cell. Really?

The only *Superintelligent Cause* known to mankind is the God of our focus nations (USA, UK, Australia, and Canada). The God of "In God We Trust," "One Nation Under God," "God Save The King," "God Keep Our Land Glorious and Free," the God of "Christmas," "Easter," and "Thanksgiving Day" -- *that* God.

He constructs our cell parts, cells, and our body using atoms from our foods which He made with the same atoms He found in the dust or soil in gardens, fields, and orchards.

"...for dust you are and to dust you will return." Genesis 3:19, NIV

"For you created my inmost being; you knit me together in my mother's womb. I praise you for I am fearfully and wonderfully made; ..." Psalms 139:13-14, NIV.

Van Wedeen and L.L. Wald of the Martinos Center for Biomedical Imaging Human Connectome Project, state that "The brain's many regions are connected by some *100,000 miles* of fibers called white matter, enough to circle the Earth four times." [1] Making our brains out of "dust" requires Superintelligence, wouldn't you agree?

Neuroscientist Harvard Professor Jeff Lichtman is studying brain composition. He was interviewed by Carl Zimmer who wrote in *National Geographic* in February 2014, "So far, the largest volume of a mouse's brain that Lichtman and his colleagues have managed to re-create is about the size of a grain of salt. Its data alone totals a hundred terabytes, the amount of data in about 25,000 high-definition movies." They also related that a mouse's brain contains about 70,000,000 neurons and a human brain contains about 1,000 times that number. [2]

Douglas Axe, PhD, is an engineer-turned-molecular-biologist. In his recent book, *"Undeniable: How Biology Confirms Our Intuition That Life Is Designed,"* he reminds us that, *"The human brain is different.... Being the most remarkable component of the human body, it is arguably the most outstanding physical invention ever to exist."* [3]

This phenomenal computer, our brain, had to be superintelligently built of atoms from the soil, air, and water with God's phenomenal two-step process: atoms in the soil are built into food, then those same atoms in our food are used to construct our brain cells.

Where else could these atoms come from, and who else has the Superintelligence to assemble them in such a phenomenal manner?

Then, of course, there is the matter of adding "Life" to these inanimate atoms. (More Superintelligence required).

" … the Lord God formed the man from the dust of the ground and breathed into his nostrils, the breath of life …". Genesis 2:7, NIV

Every living cell, including all of ours, needs God's "breath of life" today. When that is removed from any cell, it dies.

God's type of "evolution" work is shown here in deconstructing a <u>crawling</u> worm and reconstructing the same atoms into a beautiful <u>flying</u> butterfly in 14 days, not millions of years.

Stephen Russell Smith Photos/Shutterstock

Evolutionists might say this is just a "metamorphosis," but the name does not do the work of changing a crawling worm into a beautiful flying butterfly in 14 days.

Questions:

1. How do we know that it takes a Superintelligent entity to build living things, including us?

_____
_____
_____

2. Who or what is the only known entity with Superintelligence, far beyond mankind's level? _____

3. Do our governments recognize this superintelligent entity and how do they show it? _____
_____
_____
_____
_____
_____
_____

4. Where do the atoms come from to make our cell parts and cells?
_____
_____

5. According to Van Wedeen and L.L. Wald, about how many miles of fibers are made for our brains to connect the various regions?
_____

6. What does engineer-turned-molecular-biologist, Douglas Axe, believe is the most outstanding invention ever to exist and what is it made of? _____
_____

7. How long does it take our Creator to convert a crawling worm into a flying butterfly? _____

References:

[1] Wedeen, Van, and L.L. Wald, in article "Secrets of the Brain" by Carl Zimmer, *National Geographic,* February 2014, p. 34.

[2] Lichtman, Jeff, in article "Secrets of the Brain" by Carl Zimmer in *National Geographic,* February 2014, pp. 39, 43.

[3] Axe, Douglas, *Undeniable,* Harper One, New York, NY, 2016, p. 259

Chapter 3: Why Not "Evolution"?

A Shocking Admission Regarding Evolution/Materialism

One honest evolutionist, Harvard professor Richard Lewontin, explains the foundational evolutionary reasoning. His words were, *"We take the side of* (evolutionary) *science in spite of the patent absurdity of some of its constructs, in spite of its failure to fulfill many of its extravagant promises of health and life, in spite of the tolerance of the scientific community for unsubstantiated just-so stories, because we have a prior commitment, a commitment to materialism. It is not that the methods and institutions of science somehow compel us to accept a material explanation of the phenomenal world, but, on the contrary, that we are forced by our a priori adherence to material causes to create an apparatus of investigation and a set of concepts that produce material explanations, no matter how counter-intuitive, no matter how mystifying to the uninitiated. Moreover, that materialism is absolute,* **for we cannot allow a Divine Foot in the door.***"* [7] (Emphasis added).

This is both anti-God and anti-science. Such thinking disallows scientists to follow the evidence wherever it leads without reprisal.

Here Is the Simple Solution to the God vs Evolution Debate:

Either evolution is the origin and cause of life, or it is not. Proof now exists to show that Superintelligence and skills, far above the human level, are essential to construct our various cell parts and cells for our grown foods and us. Evolution, by definition, has no intelligence to use; therefore, it is *factually falsified* as both the origin and the cause of life. Evolution should no longer be taught as the origin or the cause of life as it is a lie.

Unfortunately, students will still have to say "evolution does it" on your public school and university exams for a while longer, but you can also remember the Truth of the above factors regarding how much your Creator cares and works for you.

You usually should not challenge your teachers over this fact, but you might lend them this booklet and ask for their opinion. You can decide what is logically and significantly true.

As "evolution" has many definitions, we must define the word so the author and readers can all be 'on the same page.'

The definition of "evolution" that we are using at The Atomic Biology Institute, includes the theories of evolution currently being taught in our public schools, colleges, and universities, including Darwinism, Neo-Darwinism, and macro-evolution. They include the theoretical concept of a "universal common ancestor" formed by a chance assembly of atoms billions of years ago. This unsubstantiated "universal common ancestor" had to be complex enough to have life, to function, find nourishment and digest that for survival, and to have organelles capable of reproduction including improvements with all the highly complex molecular machinery necessary to enable this work to be completed.

The mathematical odds against this just happening with no intelligent work are so enormous as to make it realistically impossible.

No one is able to adequately explain how such a complex common ancestor could have come into existence without intelligent work. In truth and reality, there is a "common Designer and Builder."

For the imagined "common ancestor" to have the intelligence and ability to assemble an offspring with even more sophisticated features, is so improbable that many, if not most, rational-thinking evolutionary scientists dropped their belief in a common ancestor.

The highly complex construction work that goes into building even the simplest living entities, is turning an increasing number of scientists away from "evolution" as the origin and cause of life.

Cambridge University Ph.D., Stephen C. Meyer, stated in his book, *Signature In The Cell*, that "The simplest extant (still surviving) cell, *Mycoplasma genitalium* – a tiny bacterium that inhabits the human urinary tract – requires 'only' 482 proteins to perform its necessary functions and 562,000 bases of DNA (just under 1,200 base pairs per gene) ... .

"Based on minimal-complexity experiments, some scientists speculate (but have not demonstrated) that a simple one-cell organism might have been able to survive with as few as 250 to 400 genes."[1]

Can anyone honestly believe that a living entity, even that small, could be constructed by chance using no intelligent work or direction? Let's be realistic and honest.

We have pointed to the Nobel Prize winners' evidence showing (unintentionally) that superintelligence is essential to build the molecular machines required for our cells to function. Evolution claims that intelligence is not required to produce life. All that is required is time, chance, and the claimed four forces that hold the universe together: gravity, electromagnetism, and the strong and weak nuclear forces.

In addition to no evidence of a common ancestor, and the evolutionists' claim that no intelligence is necessary for building cell parts and cells, are the following factors supporting a Creator:

* The atoms our cells are made of do not have the ability to move themselves into their precise position in a cell's organelles. This requires a superintelligent entity to find, select, count, and precisely assemble and fasten the atoms in the proper sequence.
* The fossil record shows no evidence of the millions of transitional forms necessary for the major categories of life to evolve. The very few fossils claimed to be transitions are dubious at best. This was a major concern for Darwin also.

*   *"Dead dogs don't bark,"* which is to say that although all the required atoms, molecules, and cells are precisely built and placed into their correct position for eyes, ears, teeth, brain, legs, heart, and so on, *without the divine "breath-of-life," those atoms, molecules and cells, or the dog, are not going to move one millimeter.* This God-given breath-of-life is crucial to every living entity. When removed, the entity's life ends.

All living entities are built of cells containing DNA. DNA is like a computer software program which needs a very sophisticated, highly complex coding programmer to arrange for the multiple and varied functions that cells have to perform in order to help keep a living entity alive. Complex, intelligent, functional codes, like those in DNA, cannot be written or programmed without an intelligent coder or programmer.

Microsoft founder, Bill Gates, stated, *"Human DNA is like a computer program but far more advanced than any software we have ever created."* [2]

Scientist I.L. Cohen said, *"At that moment when the RNA/DNA system became understood, the debate between Evolutionists and Creationists should have come to a screeching halt."* [3] He is rightly noting that when the enormous complexity of RNA and DNA programming was discovered, the theory of evolution, should have been dropped because evolution does not possess intelligence to program DNA or RNA.

Therefore, *Evolution, having no intelligence to use, is factually falsified as both the origin and the cause of life.*

Many factors exist that must be intelligently 'tuned,' highly regulated, and crucially consistent in order for our planet to function, and for living entities to exist. *Uncontrolled, unguided conditions would quickly lead to the end of all life.*

Here are just a few conditions that life must have in order to exist:

1. Temperatures must not be too hot or too cold.
2. Sufficient water must be available.
3. Sufficient food must be constantly constructed.
4. A superintelligent cause is needed to build cells from atoms.
5. God is needed to "breathe life" into these cells.
6. Adequate sunlight must exist for life to exist.
7. The atmosphere must be right.
8. Gravity must not be too weak or too strong.
9. Electricity flow must be controlled in our body.
10. Superintelligence is needed to keep all of these necessary factors (and more) in coordinated balance.

Questions:

1. According to the Theory of Evolution, a "universal common ancestor" came into being in some unknown way, with no intelligent help. What would this little entity have to be capable of doing all on its own? _____
_____
_____
_____
_____

2. Because there is no common ancestor, what is there instead that can cause living things to exist? _____
_____
3. Can atoms move themselves into their precise position in a cell?
_____
4. Does the fossil record show the millions of transition entities needed for one kind of animal to evolve into another kind of animal?_____
5. What if all the atoms and molecules for an entire animal with eyes, ears, nose, heart, brain, etc., were arranged precisely correctly

would anything else be needed for it to have life? _____

_____

6. Is any intelligence needed to assemble our DNA? _____

7. Can 'evolution' which, by definition, has no intelligence to use, program the codes necessary for our DNA? _____

8. If evolution cannot do any of the superintelligent works necessary to build a living entity, is it truly the cause of life?

_____

9. What are five of the other factors for life that require Superintelligence? _____

_____
_____
_____
_____

References:

[1] Meyer, Stephen C., *Signature in the Cell,* Harper Collins, New York, NY, 2009, p. 201.

[2] Gates, Bill, Nathan Myhrvold, and Peter Rinearson, *The Road Ahead: Completely Revised and Up-To-Date,* Penguin Books, New York, NY, 1996, p. 228.

[3] Cohen, I. L., *Darwin was Wrong-A Study in Probabilities,* Research Publications, New York, NY, 1984, p. 5.

Chapter 4: God In Our Governments.

In this book, we focus on the governments of the USA, the UK, Australia, and Canada.

Many decades ago, our forefathers and founding fathers by observation, common sense, logic, and the Word of God, understood that living entities including their livestock, food, siblings, children, and themselves, were all made by an intelligent cause whom they called "God," their "Creator," and their "Provider."

They also found truth and beneficial wisdom in His Word, the Holy Bible.

Out of respect, gratitude, and appreciation, they implemented special days of recognition for God's faithful works on their behalf including Thanksgiving Day, Christmas, and Easter.

They highly recognized God in their Declarations, Justice Systems, Constitutions, Pledges, Anthems, Mottos, and more.

They put His name on public buildings, war memorials, currency, and more.

They prayed to Him in times of personal, national, and international distress, as good leaders sincerely do today.

He is very patient and forgiving with us, but only to the point where He is purposefully disrespected or angered because praise is given somewhere else for the enormous work He performs for each of us every second of every day.

Our students have the unalienable right to be taught "Why God Is So Highly Recognized by Their Government."

History shows what happens to people and nations who turn their focus against Him.

Now, the life science of Atomic Biology gives scientific reasons that back up all governments' positions of recognizing God.

In the USA, all fifty states recognize God in their Constitutions.

Fortunately, in the USA, recognition of our Creator and appreciation of His works are regaining some prominence as states pass legislation to reinstall the national motto, "In God We Trust," back into public schools. Also, as skepticism grows against Darwinisms, some states are seeking an alternative science to teach regarding the cause of life.

It would be good to post "The Golden Rule" in classrooms as well. As Matthew 7:12 says, *"So in everything, do to others what you would have them do to you,..."*

There are challengers including the anti-God groups like the American Civil Liberties Union, the Humanists, and the Freedom From Religion Foundation. They fight against education organizations that dare to suggest there may be some intelligence used in producing living cells and entities like us. All their members would die if God quit making food for them, reliably and constantly.

As mentioned on page 9, they use a misinterpretation of the "separation of church and state." The intent of this concept is to prevent the church from telling the government what to do, and to prevent the government from telling the church what to do.

These are good concepts but we pointed out why "the church" is not God, so the two should not be confused. Also, "the government" is not "a religion."

We propose that the new science of "Atomic Biology" is a logical candidate to be considered as a suitable Darwinism alternative.

As one who lived during the "Happy Days" of the 1950's, when God and His advice were widely taught, honesty and trust were honorably practiced traits. Those of us who know the benefits of our Creator's influence, are aware of what our society has lost by the widening disrespect and ignoring of His works and blessings for each of us. Does God have the right to be angry when credit for His enormously hard work and care in providing and sustaining our lives is given to an idea like evolution which has no brain or care for life?

If we want God to bless America and other nations, we need to heed His advice and to learn about and appreciate His enormous works and care for us.

In addition to placing the national U.S. motto, "In God We Trust," into the halls of education, it would be good to explain to the students "Why" we can trust in God. Reinstalling "The Ten Commandments" at the same time would provide some lifetime wisdom and guidance for our students to live by. 'The Ten' were given to us by the God of our Governments for the benefit of all our people.

A major goal of The Atomic Biology Institute is to explain some of the multitude of reasons why we can and should trust and appreciate God and His enormous caring works for each one of us every second of every day.

A huge part of education is the teaching of Truth for practical wisdom and the long-term benefits of having and using a moral compass. Unfortunately, 'evolution' has greatly eroded the morals of our nations where it has been allowed to become dominant.[1 & 2]

Questions:

1. What four countries are we looking at where God is highly recognized by their Government? _____
_____

2. What three official national holidays that recognize God, are celebrated in these nations? _____
_____

3. What are five other ways these Governments officially recognize God? _____
_____
_____

4. Why does God deserve praise and recognition from our whole nation? _____
_____
_____
_____

5. Do you believe that God has good advice for us to use in our life and can you give an example? _____
_____

6. Regarding the "separation of church and state" concept, how do we know that God is not "the church"? _____
_____
_____
_____

7. Would you agree that God is a highly recognized part of some churches just as He is a highly recognized part of our Government?
_____
_____

References:

[1] Bergman, Jerry, *The Darwin Effect. Its Influence on Nazism, Eugenics, Racism, Communism, Capitalism & Sexism.* Master Books, Green Forest, AR. 2014.

[2] Bergman, Jerry, *How Darwinism Corrodes Morality: Darwinism, Immorality, Abortion and the Sexual Revolution*, Joshua Press, Kitchener, ON, Canada; 2017.

Chapter 5: Why Is Evolution Still Being Taught?

Evolution is still being taught because it is the enforced exclusive subject regarding the cause of life. No criticism is allowed in public education. This is truly *anti-science* because it forcibly eliminates the basic principle of scientific discovery, i.e. encouraging exploration and following the evidence wherever it leads without restriction or reprisal.

Restrictions of this type stifle the development of the best explanations, best solutions, and best education.

These 'evolution-only' rulings should be illegal, but in reality, the opposite is currently true. Teachers and professors who have dared to suggest that there may be some intelligence in the designing of living entities, have been severely reprimanded; *anti-science.*

An eminent scientist from China was speaking at a science conference in the USA and made the observation that in China you can criticize 'evolution' but not the president. But in the USA you can criticize the president but not 'evolution.'

Fortunately, now in the USA, in growing numbers, brave scientists are daring to publicly declare their skepticism of the dogma of evolution. It is bizarre that in our democracies where freedom of thought and speech have been fought for, and died for, that there should be these penalized restrictions in science, a field of such great potential for tremendous good when done honestly.

The second reason is that the religious belief in Darwinism has been allowed to dictate restrictions onto the public teaching of other beliefs. Evolution is a religious belief because it requires enormous faith to believe that such highly complex entities as living creatures can be created and sustained using no intelligence whatsoever. This is a significant conspiracy of major negative influence on our society. Evolutionists have been allowed to use our public educational institutions as their "churches" and our science teachers and professors as their "preachers" to preach exclusively their doctrine that 'evolution-only is the origin and cause of life – no God allowed.'

Six decades of enforced teaching of Darwinism as the cause of life, has blurred the image of the God of our nations, our Creator and food Provider, on whose principles our nations are founded.

History shows that ignoring God and His advice leads to trouble for the nations, including moral decay and the strife that comes with it, e.g. disrespect for law, disrespect for life, depression, anxiety, addictions, heart-ache, family breakdown, lust, greed, and crime, and all the related health-care and policing costs. All these problems come from rejecting God's wise advice, including particularly the Ten Commandments in His 'operator's manual' for all humans, the Holy Bible.

Instruction in "Bible literacy" would be of great benefit and value to **all** in our society and is being introduced in some state schools.

It stands to reason why there is less strife for people who use this great advice.

Summarizing Points:
1. You probably know this already, "Atoms Are the Building Blocks of the Universe."
2. What you have probably not been taught is that this includes us. We are also built of atoms.
3. Why were we not taught this? Because once you start looking at the enormous amount of intelligent work and care that it takes to build any one of our trillions of living cells out of atoms, "evolution" becomes *factually falsified* as both the origin and the cause of Life. Evolution, by definition, has no intelligence to use.
4. We don't like to have to say this but the Darwinian evolution that has been taught in our public schools, colleges, and universities for the last six decades as the origin and cause of life is actually a ploy to separate our students from their Creator and Provider, the God of their nation.

5. This is part of an anti-God and anti-government movement. Evolutionary professor Richard Lewontin summed up the evolution movement after listing some "patent absurdities" of some of its concepts, this way, "**.... for we cannot allow a divine foot in the door." (Bottom Line of Evolution movement).**

6. Believe it or not, this is obviously the devil's deception used to separate our students from their Creator. Public Enemy #1 has been ignored for far too long.

7. Going back to point 2, with common sense, one can ask, "(a) Where do the atoms to make us come from and (b) How are they assembled?"

8. The answer to (a) is pretty simple and obvious: our atoms have to come mainly from the food we eat and most of those atoms had to come from the dust (soil) of gardens, fields, orchards, and seas.

9. The answer to (b) is much more complex but basically obvious: our atoms have to be superintelligently, carefully, continuously, and reliably assembled every second of every day to make our foods then reassembled to make our cell parts, cells, and us..

Now that Superintelligence, far greater than that of mankind, has been 'proven' essential for the design and construction of all living cells and entities, including us, it is time to bring the God of our nations back to our students. They have the unalienable right to be taught "Why God Is So Highly Recognized By Their Governments," especially now for scientific reasons.

For those who say that no science can be "proven," this is primarily an evolutionist's statement as their theory cannot be proven. In fact and in truth, it is now proven to be false.

If you suggested to an evolutionist that he or she jump from the top of a 20-story building to show that gravity is not 'proven,' he or she would probably 'back off' (no pun intended).

The tragedy is that so many have been taught the deception that evolution is the origin and cause of life.

Misinformation taught for a long time becomes more difficult to replace as this means changing one's mind.

Questions:
1. What is the first reason 'evolution' is still being taught?
   _____
   _____

2. Why is this "anti-science"? _____
   _____
   _____
   _____

3. What is the second reason 'evolution' is still being taught?
   _____
   _____
   _____
   _____

4. What are some of the problems caused by ignoring God and His advice? _____
   _____
   _____
   _____

5. What new proof is there for the cause of life that should bring God back to our students? _____
   _____
   _____

References:

[1] See Bergman, Jerry, *Slaughter of the Dissidents: The Shocking Truth about Killing the Careers of Darwin Doubters,* Leafcutter Press, Southworth, WA, 2012, and *Censoring the Darwin Skeptics: How Belief in Evolution is Enforced by Eliminating Dissidents,* Leafcutter Press, Southworth, WA, 2016.

Chapter 6: Why Does Our Creator Care For Us?

Because our Creator's heart, mind, and capabilities are so much greater than ours, it is virtually impossible to thoroughly understand Him. However, He says He made us in His image and he obviously cares for everyone of us immensely, especially if we acknowledge Him and show our appreciation for His work for us. And amazingly, we can communicate with Him at any time.

The wisdom He provides in His Bible for us to live by, is for our most enjoyable life.

"I (Jesus) *have come that they may have life, and that they may have it more abundantly."* (John 10:10. NKJV)

We can observe His magnificent works all around us with His construction and maintenance of the people around us, our foods, our pets, and the beautiful flowers, birds, tropical fish, trees, mountains, waters, Sun, Moon, and stars we can enjoy.

Because His work for us in constructing all our grown foods and then constructing all our cells from our foods is so consistent and so reliable, we could easily take it for granted.

We can all be grateful for the enormous work He performs for each of us. He really earns our worship, praise, and thanksgiving.

He generally gives us a lifetime to acknowledge Him and His works for us, but, in the end, if we do not acknowledge Him and His Son, Jesus, and all they have done for us, He has a right to be upset and send the ungrateful souls away from His presence to a punishment place.

Less obvious Godly works for our life but equally as profound, are His essential controls of gravity, electromagnetism, the strong and weak nuclear forces, electricity, light, temperatures and the role they each play in our lives.

This God-based life science of Atomic Biology (nicknamed "God's Biology") reveals many of the details of His care for us every second of every day.

For all these awesome works done for us, there is no charge. Like the loving parent that He is, He keeps giving and giving to us, whether we thank Him or not. But, like any parent, He does expect some appreciation and recognition before we leave this world.

It is logical and understandable that He can become angry when someone or something else is given the credit for all of the constant and faithful work He performs for each of us every second of every day. The best-selling *Holy Bible* and other history books describe the troubles that come upon peoples and nations that disrespect Him and give credit for His enormous works to other things from idols to evolution.

Among the many benefits and values of knowing our awesome Creator, Provider, and Maintainer is in cherishing the 'proof' of His obvious works, love, and care for each of us.

This knowledge and understanding will be particularly beneficial for those students who learn it. The wisdom He provides in His best-selling 'operator's manual,' the Bible, has been of great benefit and value for centuries to those individuals and governments that have used it.

No scientist, teacher, or professor should be forced to pretend that there is no intelligence involved in the design and construction of living entities, including us.

One wonders how so many scientists who should know better, have been so obviously deceived or intimidated. There are, of course, many who just do not want to acknowledge or appreciate their essential, superintelligent Creator and Provider, but they should not be allowed to keep the evidence away from others, especially students.

"The wrath of God is being revealed from heaven against all the godlessness and wickedness of men who suppress the truth ..." (Romans 1:18. NKJV)

But for those of us who now know some of the many ways our creator works and cares for us, let's share this great news.

Questions:

1. In whose image are we made? _____

2. Can we communicate with our Creator? _____

3. How can we observe His magnificent works? _____

   _____

   _____

   _____

4. How much of the time does God work and care for us? ____

   _____

5. How much does He charge for all His work? _____

6. What do you think He would like from us? _____

   _____

   _____

   _____

7. Do you think God has the right to be upset and even angry with people who ignore or insult Him and show no appreciation for all the work He does for each one of us? _____

   _____

8. Where is His good advice written for us? _____

9. Do you think it is right or wrong for science teachers to be forced to teach things that they know are false? _____

Chapter 7: How Can We Individuals Use These New Facts of Life?

If you already believe in the triune God as your Creator, Provider, and Savior, you are blessed and fortunate indeed.

If you are like most believers, you probably have some friends or relatives who you would like to help find 'the way.'

A major goal of this book is to provide some solid, verifiable Truth about Life that shows how all of us are so enormously cared for and worked for every second of every day by our superintelligent Creator, the God of our nations.

All of our friends and relatives deserve the opportunity to at least hear about these works that their Creator performs for them every second.

These newly proven Facts of Life are good reasons for gratitude and thanksgiving, even at every meal, as we enjoy and get our life from the foods that God consistently and reliably continues to produce for us from His good Earth. He certainly earns our praise.

You might consider starting a book study group to discuss what is written here for understanding God's care for our life.

"God's Biology" does not seek to change the verified mechanics of the function of various cell parts, but it does seek to verify the following facts of life:
- Scientists have shown that mankind does not have enough intelligence to build any living cell out of elements, such as making living food cells out of dirt and water;
- This proves that it takes far more intelligence than mankind has and that is why we call it "Superintelligence;"
- There is only one source of such Superintelligence known to mankind, and that is our Creator, the God of our nations;
- *Evolution has no intelligence to use and is therefore factually falsified as the origin and cause of life;*
- The enforced exclusive teaching of evolution as the origin and cause of life has forced teachers and professors to teach a serious deception about life for the last sixty years;

- Teaching truth, especially about the science of life, is essential to gaining accurate results in life research;
- The phenomenal complexity of the design and construction of living cells and entities is turning honest scientists away from the lie of evolution as the origin and cause of life.

We have identified seven other principles and eighteen other essential works necessary to cause life that evolution is incapable of performing. These are described in our full study/textbook titled, *"God's Biology - Darwin's Replacement - Textbook."*

What Can We Individuals Do to Help Bring 100% of the Credit For Life, Back to Our Creator?

You might consider using this information to bring this good news about God to those who need good reasons to believe in Him.

1. If you are a pastor, youth group leader, Sunday school teacher, or Bible study leader, you can use this booklet to show God's enormous work and care for everyone every second of every day.

2. If you are a seminary leader or professor, you can use this information to teach God's enormous work and care for everyone every second of every day.

3. If you are involved with a Christian school association, you can show this information to the leaders and science teachers for their use in teaching about God's enormous work and care for all.

4. If you have children in grade school, you can explain to the principal and the life science teachers that you want the Truth about the origin and cause of life to be taught, not the proven falsehood of evolution. You can explain what you have learned here and/or lend them or give them a copy of this inexpensive booklet.

These booklets plus the full study/textbooks are available online or by contacting us at: admin@atomicbiology.com.

Question:
What other ways could you use this information? _____

Chapter 8: Choices and Consequences

We all know that there is depressing misinformation and bad news all around us these days, so are you willing to share this good news with some others?

You are probably aware of some of the many (apparently about 1430) groups who are trying to destroy our society from within. That is one of Satan's main goals and he is very good at doing this to individuals also.

What better way to bring confusion and upset to our students than to separate them from their Creator and His wisdom for their life?

We believe that many science teachers and professors would highly prefer teaching the Truth about the origin and cause of life rather than the deceptive and exclusive lie of evolution.

They need our help to stand up to the powers that enforce the teaching of the lie that separates students from their Creator.

Summarizing points:

1. Over the last seven decades, scientists have been trying to produce a living cell out of atoms and they cannot come anywhere close. This verifies that far more intelligence than mankind has is essential to build living cells and entities. We call it "Super-intelligence."

2. There is only one source of such superintelligence known to mankind and that is the triune God of our focus nations (USA, UK, Canada, and Australia).

3. These verified points *factually falsify "evolution"* as both the origin and the cause of Life.

4. We have been working full-time for the last seventeen years to prepare a basic new life science to replace "evolution" and Darwinisms. It is called "Atomic Biology" and nicknamed "God's Biology."

5. Our textbook, "God's Biology – Darwin's Replacement," and this latest summary booklet, "God Is In The Grocery Store," **are customizable to suit your group. Your additional points can be added.** Courses are coming as well and can be customized.

6. You can be a part of this significant movement to bring 100% of the credit for Life back to our Creator, the God of our nations, and to bring Him and His care and wisdom back to our students.

7. Please see the website: www.godsbiology.com for more information, contact forms, questions, and feedback.

"All that is necessary for evil to triumph is that good men [and women] do nothing." - Edmund Burke

Please choose to do something with this information to show family, friends, and neighbors these ways their creator works and cares for them. You can remind them that the Grocery Store is full of God's essential handiworks for their life.

Please get in touch with us if you would like to help expand this "Truth For Life Education Project."

God's Superintelligence Is Proven Essential For Life And Evolution Has No Intelligence To Use
IT IS THAT SIMPLE !!

WITHOUT GOD, THERE WOULD BE NO FOOD AND NO LIFE.

"God's Biology - Darwin's Replacement"
www.GodsBiology.com

About the Authors
(Each has contributed to all chapters).

Dr. Jerry Bergman is a multi-award-winning professor and author. He has taught biology, biochemistry, anatomy, genetics, psychology, and other courses over 40 years at the University of Toledo, the Medical College of Ohio, Bowling Green State University, and other colleges. His prime academic degrees include two doctorates and his total of 1,026 college credits are equivalent to almost 20 master degrees. He is one of the most formally educated people in the world.

His 1400+ publications are in both scholarly and popular science journals. Dr. Bergman's work has been translated into 13 languages including French, German, Italian, Spanish, Danish, Polish, Czech, Chinese, Arabic, and Swedish. Books that include chapters he has authored are in over 1500 college libraries in 27 countries.

To date over 80,000 copies of the 60 books and monographs he has authored or co-authored are in print.

Beyond his classroom teaching, he has been an invited speaker at many colleges, universities, and church groups in America, Canada, Europe, the South Seas Islands, and Africa.

Ramon Williams/Worldwide Photos.

Dr. Graham McLennan is one of our co-authors and history advisors regarding the role of God in our Governments. His encouragement for this project of developing "atomic biology" as a new life science has been solid since 2011.

Graham received his Degree in Dentistry from Sydney University, where he also became a Christian. He served as a Captain in the Australian Army; received the Defence Medal and the National Serviceman's Medal. Graham founded the National Alliance of Christian Leaders (NACL) with like-minded ministries in 1986.

Later, with his wife, Pam, he founded the Christian History Research Institute in 1988 and later the www.chr.org.au website when the Bicentenary Celebrations occurred and he was on the executive of the National Gathering that surrounded the New Parliament House in prayer. More people turned out for this than the opening by the Queen.

In 1993, Graham was an executive member of the Bicentennial of Christian Education, and in 2012 he initiated the National Christian Heritage Sunday celebrations. He received the Presidential Medal from the President of Vanuatu for "Services to the Nation."

He stood as a Senate candidate in 1984 and 1998 and supported others in local and state elections.

In addition to being a dental surgeon and tutor at the nearby University Dental School for many years, he also supervised dental students in Cambodia.

As convener of the NACL, Graham has been involved in helping initiate the Canberra Declaration, the Australian Christian Lobby, the National Day of Prayer and Fasting, and the Religious Freedom Institute (1990s). He is the founding Chairman of Rhema FM 103.5 and Orange Christian School. He is also a founding director of UCB's Vision FM and the Australian Christian Lobby and has served on many other national and international Christian boards and charities, as well as authoring Christian books and articles.

**Thomas Rogers** is an independent researcher and president of The Atomic Biology Institute, Lifetime Reference Guides Inc., and other companies. He has studied at three universities and two specialty institutes. His work background includes engineering, research, construction, international manufacturing, and exploration.

The education and experiences in these fields helped him in 20 years of part-time and 17 years of full-time research into understanding the superintelligent physical works with atoms required to design, construct, sustain, maintain, and repair living entities, including us.

Tom believes he has "done the time and paid the price" that might have earned him a PhD if performed under a different roof. However, the independence allowed him to think "outside-the-box" without biased restrictions and to stay focused on developing this verifiable God-based life science of "Atomic Biology" that goes a level deeper than molecular biology.

Tom has been a voluntary director of various community organizations, including the British Properties & Area Homeowners Association, the Greater Vancouver Apartment Owners Association, the West Vancouver – Howe Sound Social Credit Constituency Association, and he is a member of the Salvation Army, and Calvary Baptist Church. He has memberships in the American Scientific Affiliation, the Canadian Scientific and Christian Affiliation, the Discovery Institute/Center for Science and Culture, the Creation Science Association of B.C., the Christian Scientific Society, and the American Association for the Advancement of Science.

INDEX:
A
Amazing 1,2,6,7, 10,12,14,35
Anti-science 21,31, 33
Atomic biology 1,2,4,22,27-29,35
Atoms 6,9-11,13,15, 17-20,22-25
Axe, Douglas 18,20
B
Benefits 28,29,36
Bergman, Jerry 1,30,33,43
Burke, Edmund 41
Brain 14-16,18,20, 23,25,28
Brilliant 6,8,11,12
Butterfly 19,20 Byl, John 2
C
Care 5,6,8,13-15,21, 28,32,33,35-37,39-41
Cargo, Sharon 2
Cause-of-life 8,10,11,15,16,19,21, 22,24,26,28,31-33, 39-41
Cell-parts 6,12,13,17,20,21,23, 32,39
Common ancestor 8, 11, 22, 23, 25
Comninellis, Nick 1
Complex 1,2,6, 10-13,22-24,31,40
Construction 5,6,11, 12,22, 32,35,36

Contact Us 49
Counting 13,15,21, 23,29
Creator 2-6,8-13, 16,20,21,23,27,28,32, 35-37,39-41
Darwin, Charles 1,3,6,8,10,11,14,16, 22,23,26,28,30-33,40, 41
Design 3,7,12,18,22, 31,32,36,40,45,
DNA 13,15,22,24-26

E

Essentiality 6,8,10, 17,21,23,32,35,36, 39-41
Evolution 1,2,6,8, 10-12,14-17,19, 21-2628,29,31-33, 36,39-41
Eye 14,15,23,25
F
Feringa,B.L. 10,17
Fibers 18,20
Fossils 23,25
G
God 1-11,15,17-19,21,24,25,27-30,32,33,35-37,39-41
I
Incapable 40
Institute 22,29,44,45

Intelligent 1,2,4,6,8, 11,12,14,16-18,20, 22-27,36,39
L
Lewontin, R. 21
Lichtman, Jeff 18,20
M
Matzko, George 1
Meyer, Steve 22,26
Machines 10,12,14, 17,23
P
Pallister, C.J. 16
Perutz, Max 16
Q
Quadrillion 13,15
R
Real 11,17,22,23,31
Reber, Paul 14,16
Red blood cell 13,15
S
Sauvage, J-P. 10,17
Snoke, David 2
Speed 6
Stoddart, J.F. 10,17
Superintelligence 1, 2,4-6,8,10-12,16-21,23-26,32,36,39
T
Taylor, Jack 1
Tortora, G.J. 16
True 1,16,22,31
U
Undeniable 18,20
V
Values 36
W
Wald,l.l. 18,20
Wedeen, Van 18,20

Contact Us:

For information on our **customizable** textbooks, study books, courses, speakers, discounts, donations, participation, or other questions, please use our website: www.GodsBiology.com

As concern deepens regarding the direction our society is sliding, this concern has to become action!

For those of us who want to make a real difference for good, we beg of you to consider the information herein as something you can use to help bring the God of our nations (USA, UK, Canada, and Australia) back to our students and society.

We would appreciate your prayers, your "spreading the news," financial help for publicizing the news, your suggestions, contact info for people we should reach out to, and your help to change our world for the better.

This is part of **"The Truth for Life Education Project."**

The past six decades of teaching the deception of evolution as the origin and cause of life, has been misleading, confusing, and a destroyer of understanding regarding the enormous work and care our Creator provides for every person every second of every day.

Please see our website for further useful information.

Blessings to you and yours.

Tom Rogers, President
The Atomic Biology Institute
www.GodsBiology.com

TRUTH FOR LIFE EDUCATION

It took some time to work this out
But now we know beyond all doubt
Man does not have enough intel
To build a single living cell.

We need our superintelligent God
To grow our food and build our bod.
He's always earned the total credit
What's taught as Life's cause needs an edit.

To keep on teaching "evolution"
Gives our students harmful delusion
Separating them from their Creator
Blocking knowledge of their awesome maker.

What teacher wants their legacy
To show "They kept teaching a fallacy."
Denying their students of His great love
Is stealing their blessing He gives from above.

Through seventy years scientists have tried
To build a cell but could not guide
The correct atoms in correct numbers
Their inadequate intel, them encumbers.

Only the God of our nation can do this
Making our food from atoms into us
This significant answer can end much strife
"WITHOUT GOD, WE'D HAVE NO FOOD OR LIFE."

                                          Tom Rogers, "God's Biology"
                                          The Atomic Biology Institute
                                          www.GodsBiology.com

www.ingramcontent.com/pod-product-compliance
Lightning Source LLC
Chambersburg PA
CBHW061742070526
44585CB00024B/2772